NATURAL DISASTERS
MEETING THE CHALLENGE

HEAT WAVE AND DROUGHT READINESS

Rachel Seigel

Crabtree Publishing Company
www.crabtreebooks.com

CRABTREE
PUBLISHING COMPANY
WWW.CRABTREEBOOKS.COM

Author: Rachel Seigel

Series research and development: Janine Deschenes, Reagan Miller

Editorial director: Kathy Middleton

Editor: Ellen Rodger

Proofreader and indexer: Roseann S. Biederman

Design and photo research: Katherine Berti

Images:

Alamy
- Dennis MacDonald: p. 10
- John Wreford: p. 18

AP Photo: Gregory Bull: p. 37
Centers for Disease Control and Prevention: p. 23 (bottom)
European Commission: p. 43
Getty Images: Boston Globe: p. 30
NASA: p. 32
NASA, NOAA: p. 34 (center left)
National Center for Atmospheric Research, University Corporation for Atmospheric Research: p. 22
National Drought Mitigation Center, the U.S. Department of Agriculture, and the National Oceanic and Atmospheric Administration: p. 24
NOAA: p. 8 (bottom), 9 (bottom), 26
Our World in Data: p. 35
Shutterstock
- Capitolio Arts: p. 40
- erikjohnphotography: p. 21 (fire truck)
- EvergreenPlanet: p. 19
- Geartooth Productions: p. 6 (bottom)
- Jo Bouroch: p. 7
- MeinPhoto: p. 42
- NicoleTaklaPhotography: p. 27 (top)
- NYCStock: p. 27 (bottom)
- punghi: p 13
- rblfmr: p. 25
- txking: p. 23 (top)
- Vic Hinterlang: p. 41

United States Environmental Protection Agency, Data source: Meehl, G.A., C. Tebaldi, G. Walton, D. Easterling, and L. McDaniel. 2009: p. 6 (top)
United States Geological Survey: p. 11 (top)
Wikimedia Commons
- David Wilson: p. 33
- Library of Congress: p. 12 (inset, bottom)
- Starsend: p. 31 (both)

All other images by Shutterstock

Library and Archives Canada Cataloguing in Publication

Title: Heat wave and drought readiness / Rachel Seigel.
Names: Seigel, Rachel, author.
Description: Series statement: Natural disasters: meeting the challenge | Includes bibliographical references and index.
Identifiers: Canadiana (print) 20190231351 | Canadiana (ebook) 20190231394 | ISBN 9780778774051 (hardcover) | ISBN 9780778774099 (softcover) | ISBN 9781427125064 (HTML)
Subjects: LCSH: Droughts—Juvenile literature. | LCSH: Heat waves (Meteorology)—Juvenile literature. | LCSH: Drought management—Juvenile literature. | LCSH: Emergency management—Juvenile literature. | LCSH: Hazard mitigation—Juvenile literature.
Classification: LCC QC929.25 .S45 2020 | DDC j363.34/9297—dc23

Library of Congress Cataloging-in-Publication Data

Names: Seigel, Rachel, author.
Title: Heat wave and drought readiness / Rachel Seigel.
Other titles: Natural disasters: meeting the challenge.
Description: New York, New York : Crabtree Publishing Company, 2020. | Series: Natural disasters: meeting the challenge | Includes bibliographical references and index.
Identifiers: LCCN 2019052989 (print) | LCCN 2019052990 (ebook) | ISBN 9780778774051 (hardcover) | ISBN 9780778774099 (paperback) | ISBN 9781427125064 (ebook)
Subjects: LCSH: Heat waves (Meteorology)--Juvenile literature. | Droughts--Juvenile literature. | Climatic extremes--Juvenile literature. | Emergency management--Juvenile literature.
Classification: LCC QC981.8.A5 S38 2020 (print) | LCC QC981.8.A5 (ebook) | DDC 551.5/253--dc23
LC record available at https://lccn.loc.gov/2019052989
LC ebook record available at https://lccn.loc.gov/2019052990

Crabtree Publishing Company

www.crabtreebooks.com 1-800-387-7650

Printed in the U.S.A./042020/CG20200224

Published in Canada
Crabtree Publishing
616 Welland Ave.
St. Catharines, Ontario
L2M 5V6

Published in the United States
Crabtree Publishing
PMB 59051
350 Fifth Avenue, 59th Floor
New York, New York 10118

Published in the United Kingdom
Crabtree Publishing
Maritime House
Basin Road North, Hove
BN41 1WR

Published in Australia
Crabtree Publishing
Unit 3–5 Currumbin Court
Capalaba
QLD 4157

Contents

What are Droughts and Heat Waves?

Australia is a country of climate extremes—from extreme floods to brutal heat waves and droughts. But climate scientists have recently determined that the most recent drought, called the Millennium drought (1997–2009), was the worst in over 800 years.

That drought largely affected southern Australia, including the country's largest cities and most productive agricultural areas. Lingering remnants of this drought, as well as other droughts, led to catastrophic wildfires in 2020. The devastating blazes were fed by extreme heat waves, creating a chain of ferocious fires that tore through forests, farms, and towns.

Drought reduced Australia's Hunter River to a shallow stream through the largely agricultural Hunter Valley. The river is important for irrigating nearby farm fields.

Australian bushfires

To determine how bad the drought was, scientists studied ***ice cores,*** *coral reefs, soil, and tree rings that gave a record of climate*

400 TO 800 *years ago.*

Natural Forces Gone Awry

A heat wave is an extended period of extreme heat. Heat waves are among the deadliest weather hazards in the world. They can last for several days or weeks and are major causes of weather-related deaths in many countries of the world. According to the United States National Oceanic and Atmospheric Administration (NOAA), heat waves cause more deaths in the U.S. than any other weather-related event, including hurricanes.

Droughts are long periods of abnormally low **precipitation**, including rain and snowfall. This lack of precipitation can dry out and devastate crops and forests. Droughts put pressure on water supplies as well as water sources such as lakes and rivers, leading to water shortages. When combined with heat waves, droughts can lead to devastating fires that further destroy environments and make recovery difficult.

Natural Event to Disaster

Droughts and heat waves are natural events, but they become disasters when they impact life—killing people and animals, laying waste to landscapes and ways of life. It can be difficult to mitigate, or make a drought or heat wave less severe, but there are ways scientists, governments, and international organizations are trying to reduce threats and risks before they become disasters.

Crops and animals die during droughts. Years of drought can threaten community water supplies. Australia's droughts put 55 towns in danger of losing their water supplies to homes.

Between 1999–2009, an average of **658 PEOPLE** in the U.S. died from direct exposure to extreme heat. Between 1970–2012, drought caused close to **680,000** deaths around the world.

A Weather-Related Disaster

Heat waves and droughts are climate hazards that harm humans and the environment. A lack of rain or an extended period of heat aren't disasters on their own. When they cause deaths, illness, or severe damage to property on a mass scale, they become disasters.

When droughts and heat waves occur at the same time, the damage is made even worse. Extreme heat can make droughts more intense. Hot and dry conditions can cause wildfires, such as the bushfires in Australia, Africa, and parts of North America in 2019–2020. Long-term changes in weather patterns known as climate change have led to a major increase in **concurrent** droughts and heat waves, making their impacts more extreme.

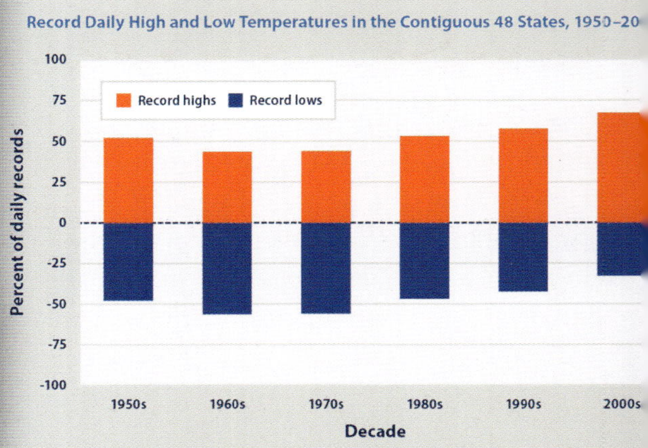

As the planet gets warmer, the chances of heat waves and droughts becoming worse increases.

Between 1990 and 2010, most of the United States saw a

20 TO 80 PERCENT INCREASE

in the number of concurrent and overlapping droughts and heat waves than in the same period between 1960–1980.

A drought fire in the hills of San Jose, California, threatens homes and ranchland in 2019.

CASE STUDY
Global Heat Wave 2003

2003 was Europe's **HOTTEST SUMMER ON RECORD** since at least 1540.

A heat wave in Europe in the summer of 2003 caused more than 35,000 deaths (some sources say 70,000) in eight Western European countries, most of which occurred over the first two weeks in August. Cities such as London, England, and Paris, France, have **temperate climates** where extreme heat is unusual. When hit with temperatures of more than 104 °Fahrenheit (40 °C) for more than eight days, governments and emergency services in these and other cities were unprepared.

From France to the Netherlands to Britain and Belgium, temperatures were 20–30 percent higher than average for that time of year, and there was not much cooling relief at night. In France, the temperature stayed at 99 °Fahrenheit (37 °C) for most of August. It was the hottest and deadliest August that had occurred in France in 150 years, and the hottest summer on record in the Northern Hemisphere.

One of the reasons the heat wave was so deadly was that many homes and public buildings did not have air conditioners. Another problem was that the **infrastructure** of cities included buildings that were hundreds of years old and were not fitted with modern cooling technologies and materials. Stone or concrete buildings held the heat at night, making it difficult for people to cool off or sleep. In France, hospitals were overrun with patients suffering from heat-related illness. Many doctors were away, as August was the traditional vacation month for doctors. The country also did not have official procedures such as a warning system in place for dealing with severe heat waves.

After the 2003 heat wave, the French government took steps to avoid a repeat of the disaster and put strict guidelines in place for handling a heat wave. These included a heat tax to fund new programs to protect the elderly, along with a heat alert system which has been used every summer since. In June and July 2019, Europe was hit with two heat waves that killed hundreds of people. France recorded its highest temperature at 115 °Fahrenheit (46.1 °C) and 1,500 deaths—10 times fewer than in 2003.

In 2019, France opened public fountains to people to cool off.

The Science Behind Heat Waves and Droughts

Simply put, heat waves happen when weather systems trap air and prevent it from moving. But it is a bit more complicated than that. A heat wave occurs when a **high pressure system** moves into an area for two or more days, forcing air from the upper levels of the atmosphere towards the ground and compressing it. These high pressure systems hover over an area, trapping the warm weather in place and making it difficult for other weather to move in.

Humid or Dry

Heat waves can be either high or low **humidity**. A high humidity heat wave is defined by continually high nighttime temperatures. In many parts of the United States, especially in the South and near the Atlantic Ocean and Gulf of Mexico, high humidity heat waves are common. Low humidity heat waves are linked with droughts. These are more common in Europe and the northern plains of North America.

Strong Winds

Some heat waves are made worse by winds blowing in from drought-affected areas. Researchers in the Netherlands determined that horizontal winds pick up heat from areas of drought and transport them to other areas of the world. This process is called advection.

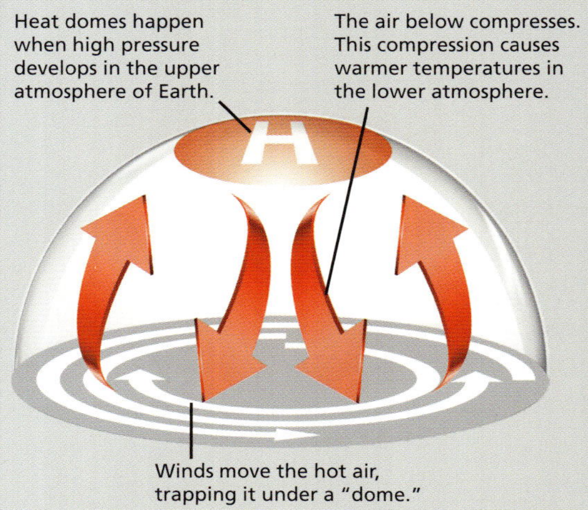

A high-pressure area that traps in heat and prevents other weather systems from moving in is called a ***heat dome****.*

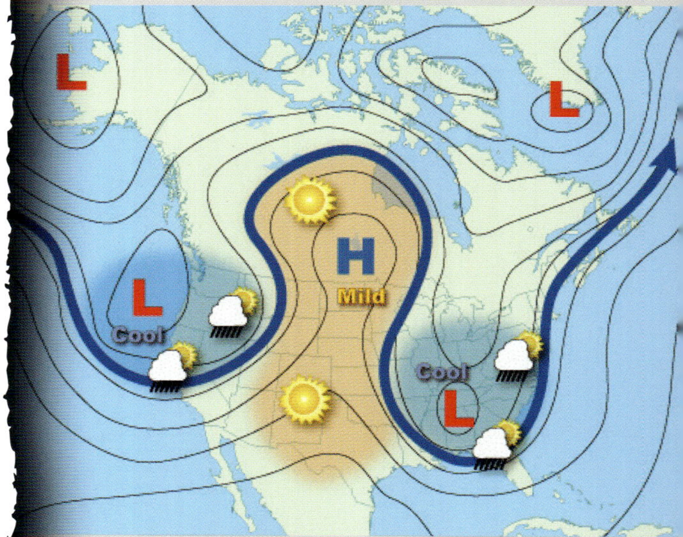

Omega Blocks are large, strong areas of high pressure with areas of low pressure on each side. They can last up to a month. The area of high pressure will have sunny and mild to hot and dry conditions.

Heat Index

Heat waves feel brutal. Heat places a strain on humans and animals. Humans have the ability to sweat to reduce body temperatures. In humid heat waves, it can be difficult to cool off because the air is hot and moist, meaning sweat doesn't offer much relief.

The Heat Index is a measure of what a temperature "feels like" when relative humidity is combined with air temperature. It is often used by weather forecasters, and is especially important during humid heat waves to warn people to alter their habits and activities. If the temperature is 90 °Fahrenheit (32.2 °C) and the relative humidity is 65 percent, the temperature is going to feel like 103 °Fahrenheit (39.4 °C). Activities such as exercising outside will put more stress on a person's body—and could lead to death—when the relative humidity and the temperature are both high.

Unlike humans, big cats and other animals cannot cool off by sweating. They pant and need to rehydrate with water.

NWS Heat Index

Temperature (°F)

Relative Humidity (%)	80	82	84	86	88	90	92	94	96	98	100	102	104	106	108	110
40	80	81	83	85	88	91	94	97	101	105	109	114	119	124	130	136
45	80	82	84	87	89	93	96	100	104	109	114	119	124	130	137	
50	81	83	85	88	91	95	99	103	108	113	118	124	131	137		
55	81	84	86	89	93	97	101	106	112	117	124	130	137			
60	82	84	88	91	95	100	105	110	116	123	129	137				
65	82	85	89	93	98	103	108	114	121	128	136					
70	83	86	90	95	100	105	112	119	126	134						
75	84	88	92	97	103	109	116	124	132							
80	84	89	94	100	106	113	121	129								
85	85	90	96	102	110	117	126	135								
90	86	91	98	105	113	122	131									
95	86	93	100	108	117	127										
100	87	95	103	112	121	132										

Likelihood of Heat Disorders with Prolonged Exposure or Strenuous Activity

Caution | Extreme Caution | Danger | Extreme Danger

Where do Heat Waves Happen?

Heat waves happen most often in hot regions, but they can occur almost anywhere in the world—even in the Arctic! They are most common in late spring through early fall but can also occur in the cooler months. People don't think as much about winter heat waves as summer heat waves because they aren't as harmful to humans, animals, or infrastructure.

Climate Change

There is no single cause of winter heat waves but climate change is a factor. Climate change is also making summer heat waves longer and more unpredictable. Scientists believe humid heat waves will increase with climate change. A study published in the scientific journal *Environmental Research Communications* in 2019 states that the number of days where the Heat Index exceeds 105 °Fahrenheit (40.5 °C) will triple between 2036–2065. The length of heat waves, or "extreme heat events," will increase too, and could double in duration.

Winter heat waves are often called **WINTER WARM SPELLS.**

Humans enjoy winter heat waves because they mean turning the heat down, extra sunshine, and not needing to wear a coat. But early thaws can cause problems with less snow melt, meaning drier soils and more droughts.

The Many Types of Drought

There are over 150 definitions of drought, so to make it easier to measure drought, scientists grouped them into four different categories.

The most common type of drought is a meteorological drought or a drought caused by longer than normal dry periods and a lack of precipitation. Meteorological drought is very specific to a region because the average amount of precipitation varies from one region to another. For example, drought in humid climates such as rainforests where it rains all of the time would be different from drought in a dry climate such as Omaha, Nebraska, where several days without rain is common.

Hydrological Drought

Hydrological drought occurs when a decrease in precipitation causes a shortage of water resources and reduces the levels of **groundwater**, **reservoir**, or stream levels. These droughts take a longer period of time to form because it takes longer for water supplies to get low, and longer to replace them.

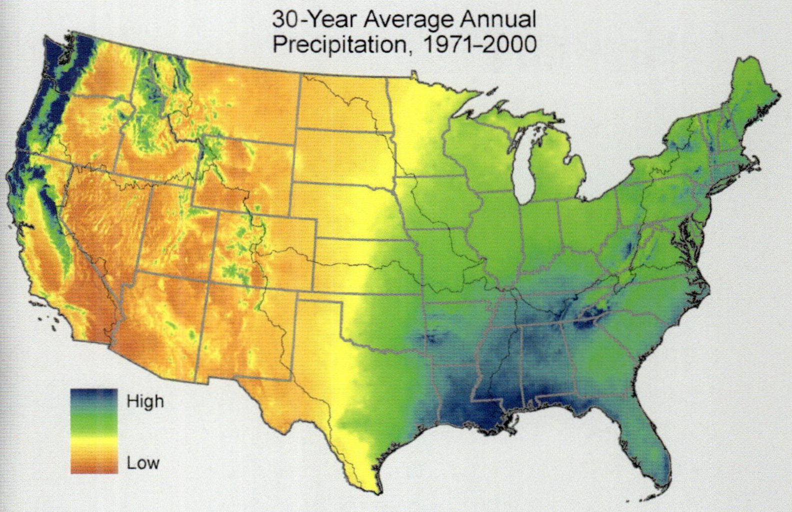

The southwestern United States has a lower average rainfall than the northern United States.

A mangrove tree in Everglades National Park, Florida, is dwarfed by drought.

Agriculture and Drought

Agricultural drought occurs when there is not enough moisture in the soil for crops to grow.

Soil conservation practices are the methods farmers use to keep the soil from losing valuable nutrients, such as rotating crops and changing locations where livestock graze. As a result of overpopulation and a shortage of farmland, there is often not enough land capable of growing crops or supporting livestock, which leads to poor conservation practices such as overfarming, overwatering, and overgrazing.

Overfarming means planting too **intensively** and not giving the soil a chance to recover, as well as overtilling, or turning the soil so much that it loses the nutrient rich upper layers.

Overgrazing occurs when livestock feed too long in a particular area. This causes the soil to press together, lowering its ability to hold water. As the soil becomes drier, it can lead to **desertification**. Overwatering uses up water supplies, making them unavailable during a drought. In the 1930s, a combination of poor soil conservation practices and natural drought led to the Dust Bowl in the United States. The Dust Bowl was a time of severe drought where winds brought blizzard-like dust storms that cleared the rich topsoil from the prairies of North America.

The Dust Bowl was considered one of the worst disasters caused by human activity and climate cycles. The prolonged drought led to 2.5 million people leaving farms on the plains of North America to find work elsewhere.

CASE STUDY

Drought and Conflict

Sometimes, droughts are made worse by war and conflict. In places where periodic drought is natural, people can survive for short periods of time using the practices they have always used. Most animals and plants can adapt. Farmers save seed or some crops for use in times of drought. In a region of eastern Africa known as the Horn of Africa, sequential drought is common. Sequential drought is drought that occurs year after year. It is being made worse by climate change. Droughts in these areas can even lead to conflict as people compete for resources such as grazing land and water. They move their livestock farther away from their traditional grazing lands into more fertile farm lands where they come into conflict with landowners and farmers.

Years of drought and civil war led to famine in South Sudan in 2017. Almost 60 percent of the population was living with severe hunger. Many were living in camps for internally displaced people.

Drought Processes

There are three physical processes that cause drought: land and sea surface temperatures, air circulation patterns, and the amount of moisture in the soil. Each of these processes is linked and if one of them changes, it can cause others to change.

As overall temperatures increase, more water evaporates, creating more severe weather conditions. The hotter it is, the higher the rate of water evaporation. Landscapes and crops need more water to survive, and water demand increases. When the moisture in the soil is used up there is less evaporation to create clouds and less precipitation. This leads to hotter temperatures which makes the soil drier.

El Niño and ENSO

El Niño is a weather pattern that occurs in the Pacific Ocean but affects weather all over the world. Every two to seven years, the ocean and wind currents off the coast of South America shift, bringing warm water into normally cold water areas. Since weather patterns are related to ocean temperatures, El Niño means more rain than usual in parts of the world and less rain than usual in other parts. Combined with the Southern Oscillation, a variation of winds and sea surface temperature in the eastern Pacific Ocean, El Niño can create what scientists call "ENSO events." ENSO is unpredictable, but often brings intense drought to Australia, India, Indonesia, the Philippines, Brazil, and east and south parts of Africa.

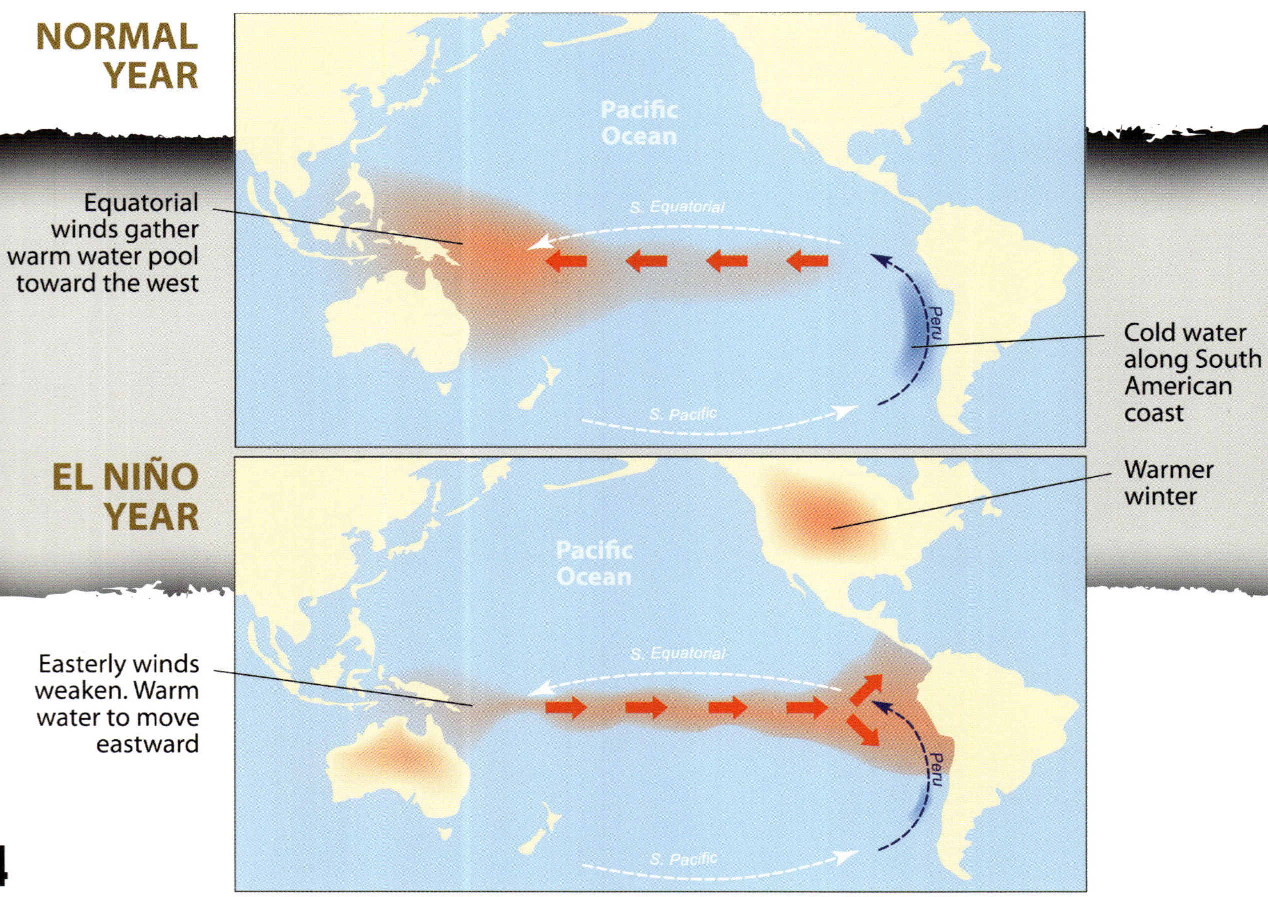

The Dry Season

Every country in the world can be affected by drought, but environmental factors make some countries more prone to drought than others. Droughts are most common in areas that have regular "dry seasons."

In the United States, droughts are most likely to occur in areas of the Southwest such as California, Arizona, Nevada, and Texas. Droughts in these areas are a natural part of the climate cycle that brings low annual precipitation and hot, dry summers.

Dry Cycles

In Canada's prairie provinces, moisture from the Pacific Ocean is carried over the Rocky Mountains where it falls as rain or snow, meaning less of it reaches the prairies of Alberta, Saskatchewan, and Manitoba. In these areas, hot summer temperatures combined with low precipitation can lead to drought. Dry weather "cycles" are increasing with climate change, but the impact of these cycles is also mitigated, or made less severe, because of changing farming practices that conserve moisture in the soil.

Outside of North America, parts of Africa, Asia, and Australia almost always experience some form of drought because of warmer temperatures and lower rainfall.

As of 2019, **ONE-QUARTER** *of the* **WORLD'S POPULATION** *living in 17 countries is experiencing* **EXTREME WATER STRESS,** *putting it at risk for severe drought.*

Arizona, USA

Alberta, Canada

Kenya, Africa

CHAPTER 2

Studying Droughts and Heat Waves

Beginning in December 2011, the State of California experienced some form of drought for 376 consecutive weeks, or seven years! This was the longest drought since the introduction of the U.S. Drought Monitor in 2000. The Drought Monitor is a series of maps and other information sources produced by the NOAA to keep track of areas of dryness and drought. In the California drought, lack of rain, extreme heat, and overcrowding of planted trees caused a mass die-off of trees in the forests of the southern and central Sierra Nevada Mountains.

Drought Emergency

In 2014–2015, California's governor made an emergency drought declaration. This declaration allowed the state to receive financial and other assistance to deal with the hardships of drought from the Federal Emergency Management Agency (FEMA). The governor ordered urban areas to reduce water use by 25 percent. Measures included encouraging people not to wash their cars, not using drinkable water to rinse off sidewalks and roads, and limiting lawn watering to two days a week.

When the Rains Came

Unusually heavy precipitation in January and February 2017 reduced the drought enough to lift water restrictions in April, but the state wasn't officially drought free until March 2019. In 2018, the governor approved new statewide standards for water conservation which include individual daily water use limits. These new laws help California prepare for future droughts and the effects of climate change.

The **2011 CALIFORNIA DROUGHT** *was the state's* **WORST DROUGHT** *in more than 1,200 years.*

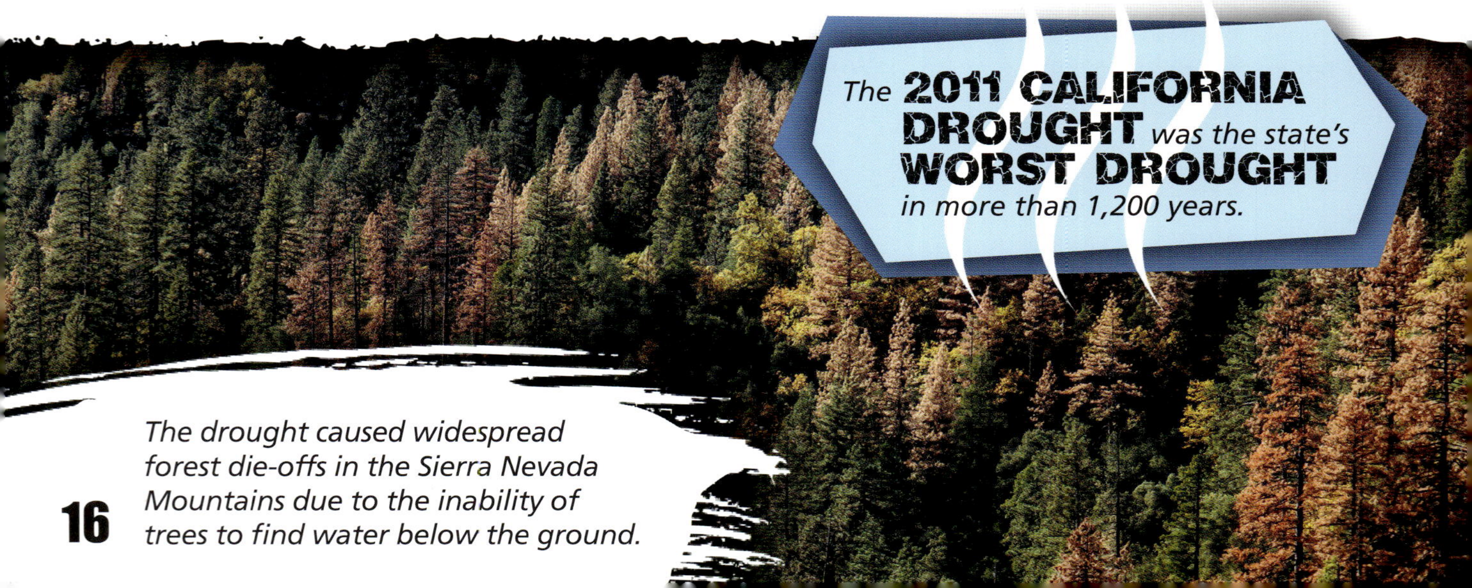

The drought caused widespread forest die-offs in the Sierra Nevada Mountains due to the inability of trees to find water below the ground.

CASE STUDY
Planning for Drought in India

In 2009, a major drought hit all of the Indian subcontinent, caused by a below average rainfall during the normally wet monsoon season. The northwest and northeast had the biggest water shortages and farmers with small land holdings were the worst affected.

In India, most farmers who depend on rainwater for crops live in areas where monsoon rain is their only source of **irrigation**. Many are subsistence farmers, growing enough food for themselves and their families. Others are mono-crop farmers, growing only one type of crop. This means they can't easily switch to a crop that needs less water. Rain-fed **agriculture** accounts for nearly half of the country's food crops, and drought affects the availability of food and water.

The 2009 drought forced the Indian government to recognize that conditions were made worse by practices such as not storing rain water in reservoirs or tanks. To address this, in 2014, it launched projects aimed in part at creating better public awareness about water conservation and water management. In 2017–18, the South Asia Drought Monitoring System (SADMS) also developed a rainfall information system that used satellite data with ground measurements of drought to improve drought response. The information was used to create maps that helped in planning for the purchase of drought-tolerant seeds and new irrigation projects. The monitoring system was also used to determine where to spray potassium nitrate, a mineral chemical that helps some crops withstand droughts.

Drought and Conflict in Syria

When drought is severe and continues for a long time, the pressure it puts on people and countries is immense. Drought made worse by climate change helped start the **civil war** in Syria. Syrian farmers were especially vulnerable to drought because their crops are dependent on rainfall and the amount and timing of rain varies from year to year.

The drought caused 75 percent of Syria's crops to fail and 85 percent of livestock to die between 2006 and 2011. This led to high unemployment and forced many Syrians to **migrate** to cities looking for work. The additional hungry and homeless in the cities strained resources. The **migrants** felt that they were being discriminated against and were not helped by the government. This became a factor in the conflict.

Health Effects of Heat Waves

Extreme heat events can have deadly effects on the human body. The body has natural systems to handle heat, but when exposed to extreme heat for too long, your heart is forced to work harder to pump blood to help your body cool off. If your body can't cool off enough, it can lead to a condition called heat stroke. Inhaling hot air can also make it harder to breathe, leading to shortness of breath.

Lengthy heat waves and droughts will make parts of the Middle East uninhabitable, which could cause conflicts over shrinking resources.

Japanese Heat Wave

In June 2018, Japan suffered a historic two-week heat wave and declared it a natural disaster. The sudden rise in temperature is blamed for causing at least 65 deaths from heat stroke in a single week, and the hospitalization of 22,647 people.

In order to prevent further heat-related deaths and injuries, the Japanese government began increasing its efforts to warn people about extreme heat. They launched school campaigns, and a new smartphone app. A new update for visually impared people was also included in their Wet Bulb Globe Temperature (WBGT) index which is a measure of how hot it feels when you factor in humidity and wind.

Extreme Heat and Infrastructure

Heat waves not only have an impact on human health, but on infrastructure such as roads and train tracks. During a heat wave in the United Kingdom in 2019, temperatures reached 100.4 °Fahrenheit (38 °C) in the East Midlands, an area of central England. It was so hot that roads began to melt. The metal on train tracks expanded and bent under the pressure.

Some people in Japanese cities take part in uchimizu, a tradition that involves sprinkling cool water on hot pavement to cool things off.

During heat waves, temperatures on road surfaces can reach up to 122 °Fahrenheit (50 °C).

Roads and Rails

In extreme heat, road surfaces absorb a lot of heat, causing them to melt and bubble like chocolate. As they harden again, they become fragile and break. Trucks normally used for snow removal are sent out to spread salt and sand to improve grip on the melting roads. Rail tracks also absorb heat and when their temperature rises to 68 °Fahrenheit (20 °C) above the surrounding air temperature, the trains are forced to slow down to avoid buckling.

How do Scientists Study Drought?

Scientists study drought by looking at **geological** records of historic droughts. For example, trees can live for hundreds or thousands of years, and over that time, they "record" different kinds of weather conditions in their growth rings. Basically, the rings on a tree not only tell us how old the tree is, but what the weather was like each year. In warm, wet years, tree rings grow wider, while in cold and dry years they are thinner. If a tree has experienced drought, it might not have grown much during those years.

Measuring Heat

The Heat Index (HI) and Excess Heat Factor (EHF) are measurements that go beyond temperature to account for the human body's response to heat. HI is a measure of how hot it really feels with the humidity. EHF uses meteorological data to provide a way of predicting deaths from extreme heat.

Paleoclimatology is the study of climates of the ancient past.

SCIENCE BIO
National Drought Mitigation Center

The National Drought Mitigation Center was formed in 1995 to reduce the impact of drought on people, the environment, and the economy by emphasizing drought readiness and risk management rather than disaster management. They are based in the School of Natural Resources at the University of Nebraska–Lincoln in the United States.

Their mission is mostly national, but researchers and staff help build resilience to drought through monitoring and planning. Drought resilience means learning how to bounce back from and adapt well to the impacts of drought. Building resilience includes learning from experience and preparing for drought instead of responding to it after it happens. They work at all levels—from individual farms to communities, states, and countries around the world.

One of the ways the Center helps communities plan for drought is through the use of drought scenario exercises. In the same way your school prepares students for a real fire with fire drills, the drought scenarios create exercises and games that use the real conditions and events that may happen during a drought. They are a useful way to bring people together and help them come up with ways to warn the public about drought, identify gaps, and develop strategies for mitigation.

Impacts in the Drought Impact Reporter

01/01/2010 to 12/31/2019

In July 2005, the National Drought Impact Center launched the Drought Impact Reporter (DIR) as the first complete database of drought impacts in the United States.

CHAPTER 3

Meeting the Challenge

Heat waves can be predicted pretty accurately a few days or a week before they happen thanks to super computers and numerical forecasting, a formula that uses mathematical equations to predict the weather. Even with these tools, long-term forecasts are difficult. Computer models can give a picture of what the weather is most likely to be like, but measuring the atmosphere that far out has limited accuracy because it is unpredictable and always changing.

Discovering New Patterns

Recently, scientists discovered a unique pattern of water temperatures across the North Pacific Ocean called the "Pacific Extreme Pattern." They think this pattern will help them predict heat waves up to two months in advance. The pattern is an area of the ocean where there is a clear difference between warmer-than-average water and cooler-than-average water. When the pattern appears, the odds of a heat wave occurring in a particular week or on a particular day can triple.

The pattern is important because most long-term seasonal weather forecasts don't focus on heat waves or other extreme events, since the data that helps forecasters predict these events isn't available months in advance.

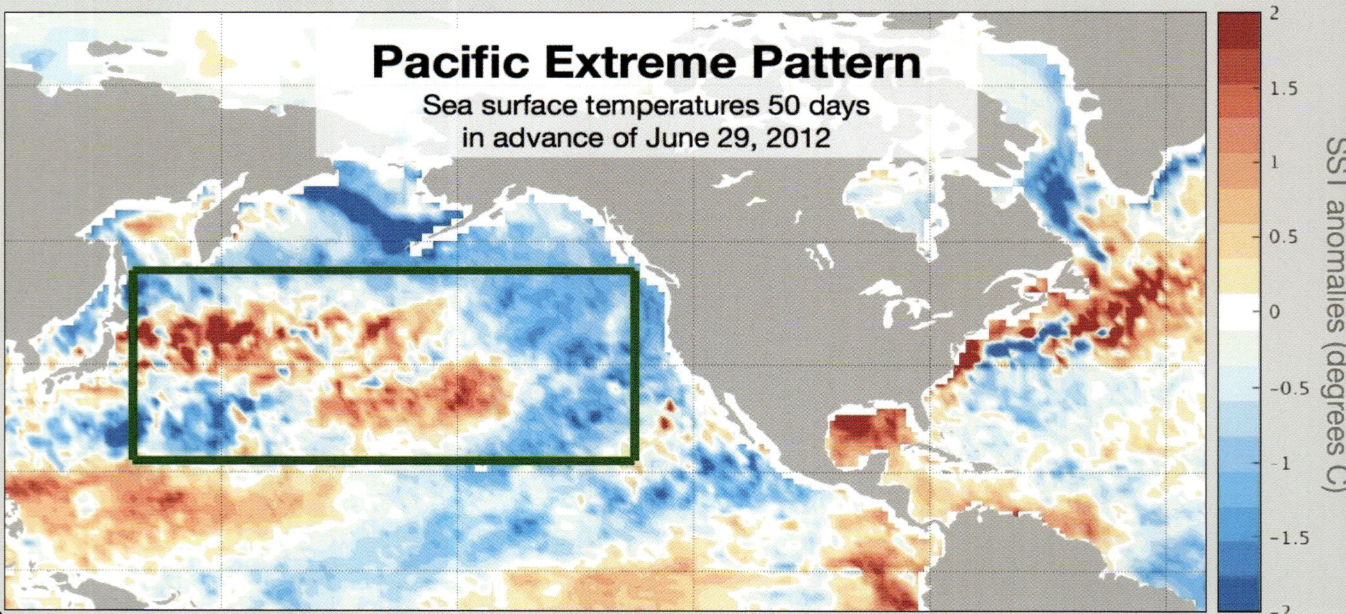

An accurate heat wave prediction can usually only be made a maximum of 10 days in advance.

HEAT-RELATED ILLNESSES

WHAT TO LOOK FOR	WHAT TO DO
HEAT STROKE	
• High body temperature (103°F or higher) • Hot, red, dry, or damp skin • Fast, strong pulse • Headache • Dizziness • Nausea • Confusion • Losing consciousness (passing out)	• Call 911 right away-heat stroke is a medical emergency • Move the person to a cooler place • Help lower the person's temperature with cool cloths or a cool bath • Do not give the person anything to drink
HEAT EXHAUSTION	
• Heavy sweating • Cold, pale, and clammy skin • Fast, weak pulse • Nausea or vomiting • Muscle cramps • Tiredness or weakness • Dizziness • Headache • Fainting (passing out)	• Move to a cool place • Loosen your clothes • Put cool, wet cloths on your body or take a cool bath • Sip water **Get medical help right away if:** • You are throwing up • Your symptoms get worse • Your symptoms last longer than 1 hour
HEAT CRAMPS	
• Heavy sweating during intense exercise • Muscle pain or spasms	• Stop physical activity and move to a cool place • Drink water or a sports drink • Wait for cramps to go away before you do any more physical activity **Get medical help right away if:** • Cramps last longer than 1 hour • You're on a low-sodium diet • You have heart problems
SUNBURN	
• Painful, red, and warm skin • Blisters on the skin	• Stay out of the sun until your sunburn heals • Put cool cloths on sunburned areas or take a cool bath • Put moisturizing lotion on sunburned areas • Do not break blisters
HEAT RASH	
• Red clusters of small blisters that look like pimples on the skin (usually on the neck, chest, groin, or in elbow creases)	• Stay in a cool, dry place • Keep the rash dry • Use powder (like baby powder) to soothe the rash

CDC

Tracking Heat Waves

People at home can monitor and track heat waves through the Centers for Disease Control and Prevention's national environmental health tracking network. The CDC is the leading public health institute of the United States, whose goal is protecting public health safety by monitoring and preventing disease outbreaks, injury prevention and control, and environmental hazards.

The tracking network brings together health and environmental data at a national, state, and city level that is hard to find anywhere else. It provides people with supporting information to make it easier to understand. People can use the tracking network to view interactive maps, tables, and charts as well as to find information by location, and view state and local tracking websites. Public health professionals use the data to find patterns of risk, plan public health programs and activities, prepare for and respond to emergencies, and to measure how well their response worked.

The tracking network provides information to help people protect themselves from heat-related deaths or illnesses.

Monitoring and Tracking Drought

Droughts can have devastating and widespread impacts on societies, which is why understanding and monitoring them is crucial. In order to monitor drought, scientists use drought indicators which are pointers that describe drought conditions such as rain and temperature, and **indexes** such as the number of days without precipitation.

The Standard Precipitation Index (SPI) is an example of a drought index. It uses the average rainfall over a long-term period of 30 or more years to create an early warning scale. When the SPI is continually negative for a period of two to three weeks, a drought begins. When the number turns positive, the drought is over.

The U.S. Drought Monitor is a weekly map that uses data gathered by drought experts to show the location and severity of drought across the country. The map classifies drought into five categories using bright colors ranging from yellow to dark red to show how severe the drought is. The categories are abnormally dry, moderate, severe, extreme, and exceptional. The experts take into account indicators such as recent precipitation totals, soil moisture, and water levels in lakes and streams, to make judgements about drought conditions. The Canadian Drought Monitor, produced by Agriculture and Agri-Food Canada, is Canada's official source for drought reporting and monitoring.

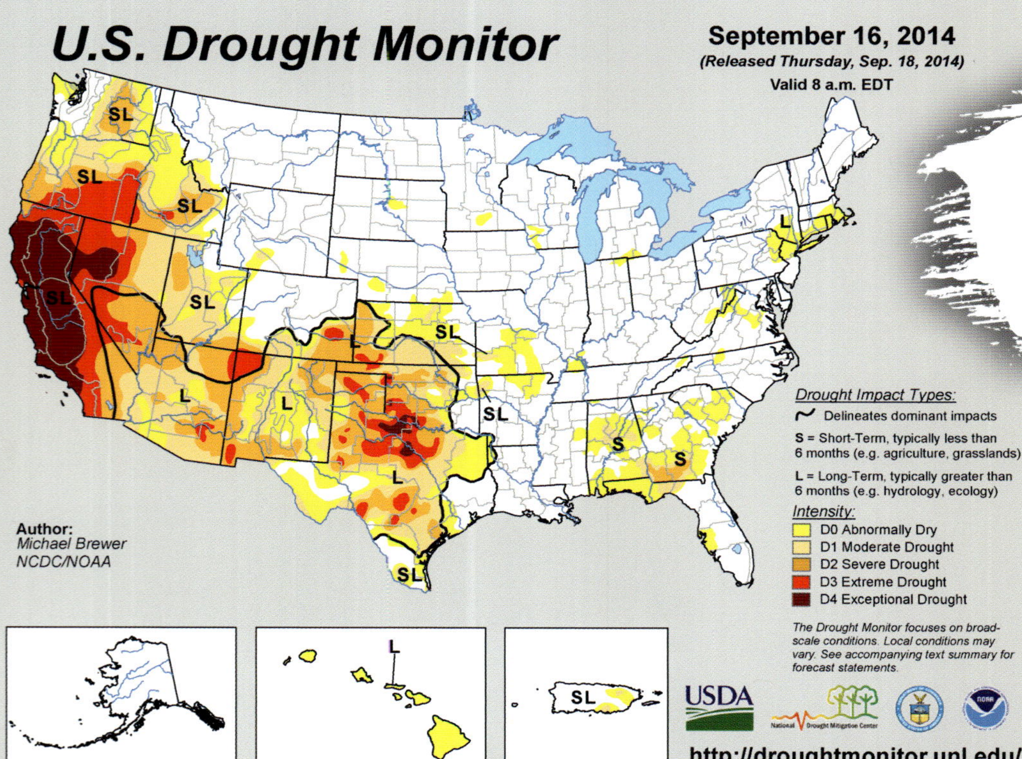

The Drought Monitor map gives an overview of drought conditions throughout the U.S. for each week.

Category	Description	Possible Impacts
D0	*Abnormally dry*	• *Going into drought: short-term dryness slowing planting, growth of crops or pastures* • *Coming out of drought: some lingering water deficits, pastures or crops not fully recovered*
D1	*Moderate drought*	• *Some damage to crops, pastures* • *Streams, reservoirs, or wells low, some water shortages developing or imminent* • *Voluntary water-use restrictions requested*
D2	*Severe drought*	• *Crop or pasture losses likely* • *Water shortages common* • *Water restrictions imposed*
D3	*Extreme drought*	• *Major crop/pasture loss* • *Widespread water shortages or restrictions*
D4	*Exceptional drought*	• *Exceptional and widespread crop/pasture losses* • *Shortages of water in reservoirs, streams, and wells creating water emergencies*

Drought is classified into five categories ranging from abnormally dry to exceptional drought.

Heat Warning Systems

The National Weather Service Forecast Office in the United States has different warning levels for extreme heat, which are based on Heat Index Values. The Heat Index is a measure of how hot it really feels when humidity is factored in. Humidity makes the weather feel hotter because our skin uses the air around us to get rid of sweat and cool off. If the humidity is high, there is already a lot of moisture in the air, and sweat doesn't evaporate. This makes us feel hotter.

195.7 MILLION PEOPLE in the United States were under excessive heat watch or warning or heat advisory on July 28, 2019.

The humidity index (humidex) is used by Canadian meteorologists and combines the effects of heat and humidity to describe how hot the weather feels.

Heat Alerts

An Excessive Heat Outlook warns people when a heat event is possible in the next three to seven days. An Excessive Heat Warning and Heat Advisory are issued within 12 hours of arrival of dangerous heat conditions. Excessive heat watches are issued one to three days before a heat wave is expected but exact date isn't known. Canada's Heat Alert Response System (HARS) was created to help protect citizens during a heat wave. HARS has three alert levels to describe extreme heat that are based on intensity, length, and nighttime exposure. A level 1 alert means people with pre-existing breathing and heart problems can be affected while level 3 means everybody will be affected.

Drought Warning Systems

Early drought warning systems are designed to warn local communities about drought risk, as well as to help improve their preparedness and minimize impact. In order to warn people about upcoming or existing droughts, the www.drought.gov website in the U.S. provides different drought reports. The National Weather Service's Climate Prediction Center publishes a monthly drought outlook for the following month for all 50 states and Puerto Rico.

The Research Institute for Climate and Society makes monthly worldwide predictions for temperature and precipitation. Users can also generate maps to show probability of precipitation and amount that will start or end a drought. The Climate Prediction Center is a U.S. agency that makes seasonal drought outlook predictions and real-time climate monitoring.

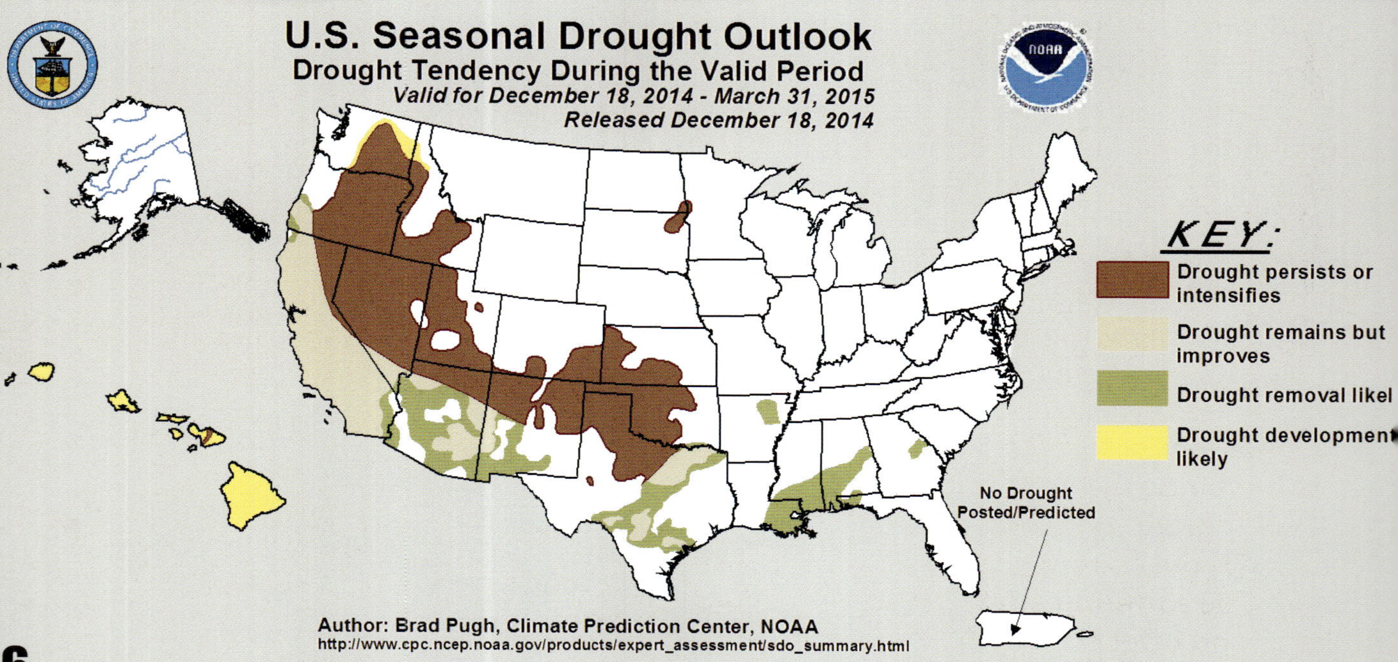

Increasing Resiliency

The impact of heat waves is made worse in cities by the urban island effect. An urban heat island is created by closely packed buildings and paved surfaces in urban areas which absorb more heat in the day and release heat at night. This makes them get hotter and stay hotter than rural areas. In order to increase **resiliency** to heat waves, cities have different adaptation strategies.

Reaching Out

Big cities such as New York City are highly vulnerable to heat waves. They are built up and paved in concrete—a material that absorbs and retains heat. As a result they have taken actions that will build resiliency and reduce the risk from heat waves. Steps include an increased use of air-conditioned public spaces known as cooling centers. These allow people who don't have air conditioning to cool off in public spaces such as libraries, community centers, schools, and malls. They also promote resiliency by sharing life-saving information with vulnerable populations such as the elderly with their Be-a-Buddy Program. This program trains volunteers to reach out to seniors who might be suffering alone in the heat, and offer them help.

Splash pads act as cooling stations in cities where concrete buildings, roads, sidewalks, and parking lots create heat islands.

Heat Wave Guidelines

In 2019, the Red Cross/Red Crescent released its first set of international guidelines on heat waves for cities. The guide offers cities practical steps to prepare for heat waves, and aims to prevent illness and death. Steps range from leading public awareness campaigns to developing early warning systems.

Limiting Waste

The most common response to drought involves limiting or forbidding water waste. This includes not leaving the water running when we are brushing our teeth or doing dishes, or leaving the hose running when washing cars. In Texas, where droughts are common, water conservation has become part of the state water plan, which offers suggestions on saving water.

The Northeast Blackout of 2003 left **45 MILLION PEOPLE** *in the United States without power during a heat wave.*

Extreme heat can put a strain on power grids. When it's hot, use of air conditioners increases the demand for electricity. This can lead to power outages and infrastructure damage.

Cape Town Water Crisis

Many cities throughout the world are at risk of running out of drinking water. And the main reason is severe drought. Cape Town, South Africa, is one of those cities. In 2018, three years of little rainfall forced one of South Africa's most prominent cities close to "day zero." This was the day when Cape Town would cut off all city water running to homes and businesses.

Water Conservation

The city forced people into severe water conservation. Water use was monitored. People were urged not to shower for longer than two minutes, told not to flush toilets unless absolutely necessary, and were given fines for using more water than the maximum of 13 gallons (50 L) allowed. Luckily, the water conservation measures and the return of rainfall helped Cape Town end the water crisis without a tap shut off.

Changing Habits

The city gradually lowered its water restrictions, but had learned it could cut peak water usage in half by forcing people to change their water use habits. The city's dam levels remain at 50 percent capacity—that's below pre-drought years. Cape Town is now following a long-term plan to diversify its water sources. The plan includes a possible **desalination** plant to make ocean water into drinking water.

Cape Town's climate means wet winters and dry summers, but years of drought and a growing population puts more pressure on water supplies.

The average U.S. household uses between
19.1 *and* **26.7 GALLONS**
(68.5 to 101.1 L) of faucet water a day.

Gaps in Heat Wave Mitigation

Many countries use heat wave warning systems, but there are still gaps in measuring risk. Most heat wave warning systems only consider environmental factors, the temperature outside, and the humidity. They don't account for how warm it is inside buildings, how healthy the person is, what they are wearing, or how well an individual copes with heat.

How People Respond

Another gap in reducing the impact of heat waves is knowing how people will respond to a heat wave. Vulnerable groups such as the elderly, young children, people living in poverty, and those with disabilities don't always have the ability to protect themselves from the impacts of heat waves. Their bodies may not cope as well, or they may not have access to cooling air conditioning. Programs that distribute water to the homeless, provide sun protection, hand fans, or cooling centers, are a part of risk reduction. They save lives and money, by preventing people from developing heat stroke or having to go to the hospital.

Some schools in the United States don't have air conditioning, even in areas that see high temperatures during the school year. Fans can help move the air in the stifling heat.

SCIENCE BIO
Oasys Water

Oasys Water is a U.S. water technology company that turns saltwater from oceans into drinking water. One of the biggest problems with drought isn't that the world has no water, but that less than one percent of the water is safe to drink or use for agriculture.

Existing methods of **desalination** require so much energy and are so expensive that many countries can't afford it.

Oasys uses a technology called forward osmosis to separate the salt from the water by passing the liquid through a **membrane** with a concentrated mixture that blocks and then heats the water to dissolve it. The method produces water in a faster, cheaper, and more energy efficient way. It is currently being used in places where there is drought and not enough fresh water.

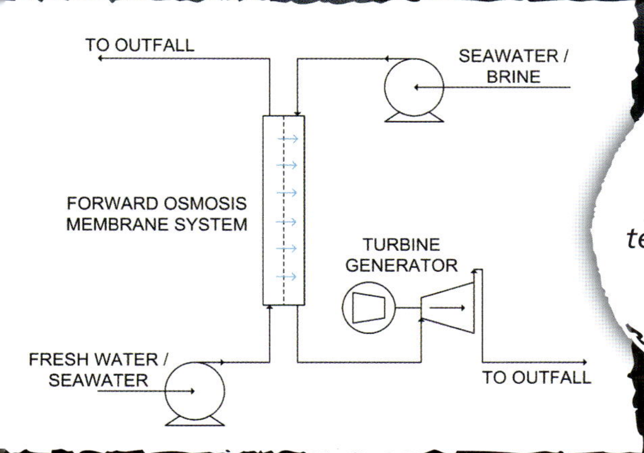

The Oasys desalination process uses one tenth of the energy required for a regular plant.

Prediction Technologies

Data gathering technologies are making it possible to more precisely predict droughts and heat waves. Some of these technologies are being spearheaded by private companies that are developing software and programs that analyze climate data and look for patterns. They use jet stream position, air temperature, thunderstorm activity, pressure patterns, and decades of weather records to predict the likelihood of a heat wave.

New Research

New technology developed by the space science industry can predict droughts and fire risks up to five months before they happen. Researchers gathered information from a pair of NASA/German satellites in space to accurately measure the water below Earth's surface. The satellites were able to supply measurements of changes in water stored anywhere on Earth, which will lead to earlier drought forecasts, allowing communities more time to manage its impact.

CASE STUDY
Chicago Heat Wave 1995

In the Chicago heat wave of 1995, record-breaking temperatures combined with record humidity caused the deaths of 739 people over a period of five days. The 106 °Fahrenheit (41 °C) temperature on July 13, 1995 set the record for the hottest July temperature since records began in 1928. In a place nicknamed "the windy city" there was also no wind to offer relief. Most of the dead were poor and elderly people who lived in the inner city and didn't have air conditioning. Many were afraid to open windows or sleep outside for fear of crime.

One of the factors that made the heat wave so deadly was that Chicago is an urban heat island. This raised the temperatures in the city by more than 3.6 °Fahrenheit (2 °C). Other factors were not having a proper warning system, power outages caused by a power grid that couldn't meet the demand, and hospitals and emergency services that were completely overwhelmed.

That week completely changed the way that Chicago responds to and prepares for emergencies. After the heat wave, the city's office of emergency management and communications placed all departments under one roof to be better organized and send resources where they are most needed. They also planted more trees on parkways, near paved parking lots, and loading docks, and installed green roofs and rain barrels to store water to reduce heat. These solutions helped reduce the number of deaths during the next heat wave.

During the Chicago heat wave, temperatures went as high as **110 °FAHRENHEIT** *(43.33 °C) and reached 119 °Fahrenheit (48.33 °C) on the Heat Index.*

Without wind to move the air, temperatures got even hotter and the insides of homes were like ovens.

CHAPTER 4

Facing Future Droughts and Heat Waves

Between 2000 and 2019, a new global temperature record has been set five times. June temperatures in 2019 broke records in Europe, Africa, and South America, and 2019 was the hottest first half of the year in Alaska, Mexico, Western Canada, New Zealand, Madagascar, and eastern Asia.

These records are connected to anthropogenic climate change, or changes caused by human activities such as increased burning of fossil fuels. Heat waves are getting hotter and are lasting longer. In a normal summer, below and above average temperatures would balance out. Over the past 10 years, daily record high temperatures have occurred twice as often as record lows across the United States, and it's staying hot longer.

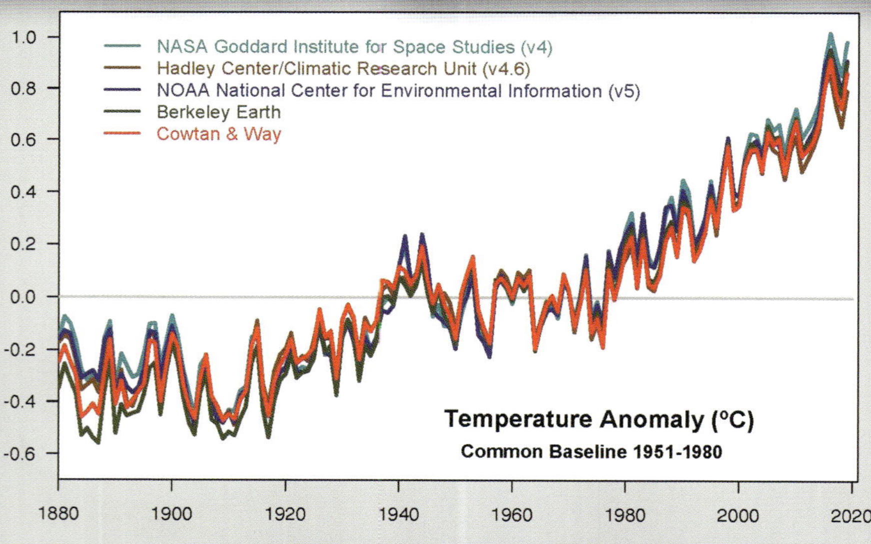

By 2100, it is estimated that **75 PERCENT** of the world's population will face deadly heat waves.

Coal-burning power plants give off a lot of carbon dioxide.

Water Use and Conservation

In most of the world, water is a finite resource—it has its limits. We use, and waste, a lot of water every day. The average American family uses 300 gallons (1,135 L) of water per day at home for bathing, washing clothing, and household cleaning. The average Canadian uses 86 gallons (329 L) per person every day. Studies show Americans and Canadians use twice as much water per day as Europeans. Our industries also use a lot of water. Population growth, agriculture, and industry increase water use and demand year by year. The recent drought in California was partly caused by the overuse and wasting of water, as well as outdated ways of managing and reporting water use.

Drought and Human Activity

Most current drought prediction models focus on rainfall, river flow, and soil moisture, but don't take into account human activities that cause hydrological drought. When combined, the picture is more complete: the effects of some droughts, such as the one that affected Cape Town's water supply, can be made less grim if human water consumption and use is reduced through conservation.

Agriculture accounts for as much as
80 PERCENT
of all water use in California

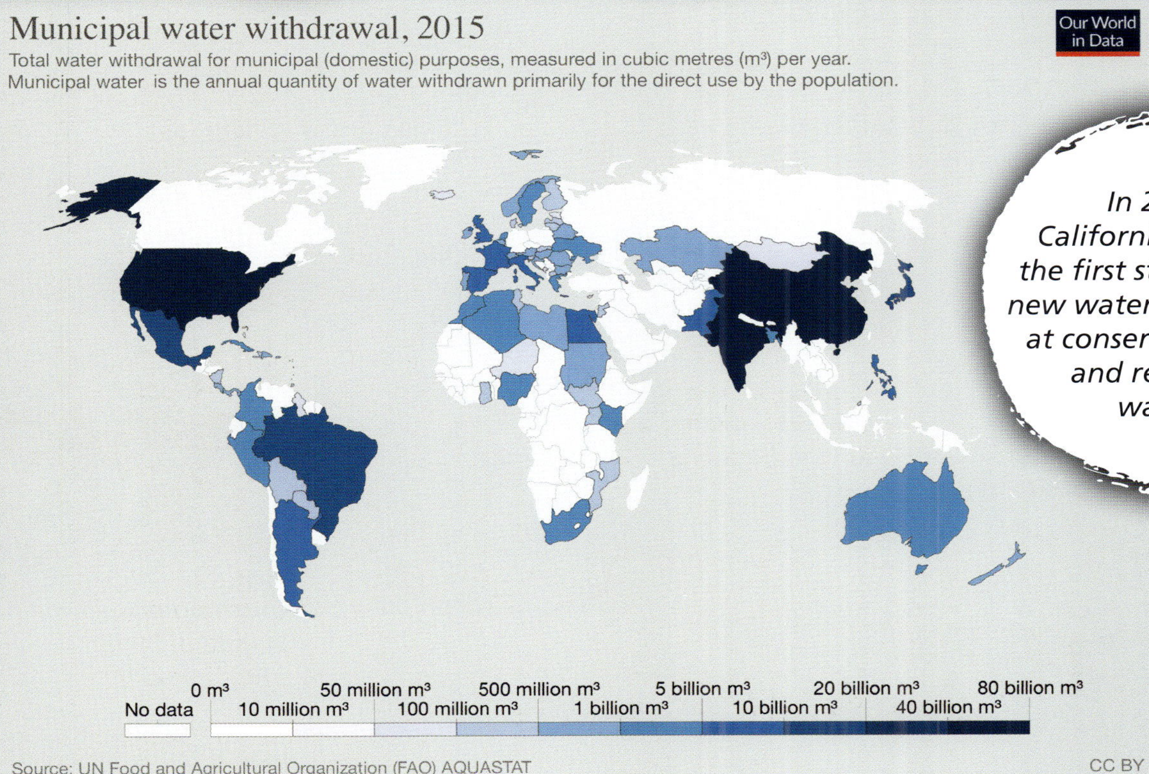

In 2018, California became the first state to pass new water laws aimed at conserving water and reducing waste.

Reducing Heat Wave Impacts

There are a number of ways that cities can reduce the impact of heat waves, but one of the best ways is through planting more trees and building more green spaces into cities. Trees, green roofs, and plants cool the air, land, and water. Shade and moisture reduces the heat in the air. This not only helps people cool off, but it saves energy by reducing the demand for air conditioning. Trees and green roofs also soak up carbon dioxide during **photosynthesis** which helps clean the air and reduce air pollution.

Replacing dark roofs with light-colored, reflective roofs in urban areas can lower temperatures and save lives during heat waves.

City Development

Cities such as Dallas, Texas, are using health, environmental, social, and economic factors to determine the best locations to plant new trees to help cool down the city. In Seattle, the construction industry is being encouraged to add rooftop gardens to new buildings, or to cover the walls with **vegetation**.

Building Green

In Singapore, an island city-state in Southeast Asia, "green" building has been a required part of construction since 2008. Green building requires that carbon converting plants be incorporated into new building construction. All new developments must include green roofs, green balconies, or walls of greenery on the outside of the building.

Using irregular shapes in buildings can create more shade on the building's face and reduce the temperature inside.

Changing Climate

With the changing climate causing droughts to be more frequent in many parts of the world, cities are looking at different strategies to reduce their impact, including water conservation. One way to do this is to stop using drinking water for irrigation or watering lawns. Paris, France, uses a dual water system which provides drinkable water to houses for people to use, and non-treated water for cleaning streets and watering parks.

Agriculture and Drought

Agricultural drought can be minimized by using smart irrigation techniques. Smart irrigation technology can cut down on water waste by monitoring weather, soil conditions, evaporation, and plant water to deliver a set amount of water to specific crops for a set amount of time.

In Santa Fe, New Mexico, developers have to provide the city with an estimate of how much water their buildings will need before they can build. Whether or not they are allowed to build depends on the amount of water available. If there isn't enough water, no new buildings are allowed. Qingdao, China, has constructed a large water desalination plant to filter ocean water and produce enough water to meet their water needs.

This reverse osmosis desalination plant in Carlsbad, California, is the largest in the U.S. It produces 50 million gallons (189,270,589 L) of desalinated water per day for household use by making ocean water into drinking water.

Awareness and Education

Practical steps called drought management options can be taken immediately to build greater resilience to drought or reduce the impact of drought. Increasing drought awareness allows communities to better understand how drought will impact their food and water supplies, and reduce shortages. Investing in early warning systems provides information early enough for communities to take action to avoid or reduce the risk of drought, and to prepare a response. A sharing of success stories and experiences among countries will aid in the development of a global drought response that will benefit poorer countries the most.

New Tech

In many drought-affected cities, low levels in their water reservoirs are forcing **hydroelectric** plants to close or lower their energy production. In order to combat this, cities such as Portland, Oregon, have started using Lucid Energy Technology, which installs rotating mechanical devices called turbines in the city's water pipes, and generates electricity as the water passes through. The city of Santa Fe, New Mexico, stores excess rainfall and surface water in an underground aquifer where it will not evaporate in dry desert air.

Use rain barrel water for your garden

Turn off the tap when brushing teeth

Don't water the lawn in the summer

Fill up your glass half-way

In 2016, Conaway Ranch in California became the first farm in the state to use drip irrigation technology to feed their rice paddies.

SCIENCE BIO

Simcha Blass, Netafim

Drip irrigation is an ancient technology used at least since 100 BCE. Ancient peoples used to bury unglazed clay pots of water and plant crops around the pots. Water seeped through the pots to the surrounding soil to water plant roots. In the 1800s, German farmers experimented with modern drip irrigation that used clay pipes to combine drainage and irrigation. Over the years, other innovations, such as plastic pipes with tiny holes, were used.

In 1965, Israeli inventor Simcha Blass developed a modern drip irrigation system that transformed dry climate agriculture. An engineer by training, Blass determined that a slow and balanced drip led to better plant growth, and he created a drip-based tube that slowly released water where it would be most effective. Partnering with a kibbutz, or collective farm in the Negev desert, Blass formed Netafim. The company grew to become the largest provider of drip irrigation systems in the world.

Blass's commercial dripper is an efficient water and nutrient delivery system for crops, allowing farmers to produce more crops with less water. Today, the technology is used by farmers in arid and drought-prone areas.

Preparing for Future Heat Waves

Heat action plans are a set of rules and regulations that are developed to protect communities from the threats posed by rising temperatures and extreme heat. Many are developed by individual cities, but with the cooperation of higher levels of government.

Indian Plan

In Ahmedabad, India, heat-related deaths have fallen about 20 percent thanks to the application of their Heat Action Plan. The plan has four primary strategies, including building public awareness of risks, early warning, promoting adaptive measures, and heat response from health care providers. Ahmedabad is a modern city of 5.6 million people in central India. Acknowledging that climate change is increasing temperatures, the city decided that it must put a heat action plan in place to prevent deaths like the 1,344 it experienced during a 2010 heat wave. It worked with Indian public health organizations to develop a plan.

The public awareness strategy of the plan included community outreach, where social media would be used to warn people of heat dangers. The plan put local medical schools, hospitals, and public health officials together to set up practices to ensure access to drinking water and cooling spaces for people most at risk: the elderly and the poor. Ahmedabad also introduced a color-coded heat alert from yellow to red, with red being an extreme heat alert. The city's Heat Action Plan was the first in the country, and has been used as a blueprint for others. It is also updated regularly.

Cities can also adapt to future heat waves by adopting the "green, blue, and cool" strategy. This involves introducing more green infrastructure by planting more trees and plants, blue infrastructure such as pools and ponds, and cool infrastructure such as shade spots and access to cool buildings.

CASE STUDY
Drought and Human Migrants

Years of drought in Central American countries have forced millions of people to migrate from their homes in Guatemala, El Salvador, Honduras, and Nicaragua. Some have gone to find work in the cities, while others have become **climate migrants** seeking refuge in Mexico and the United States.

From 2014 to 2019, drought has ruined two main food crops of corn and beans in Central America. Over 1.4 million people have required food assistance from international aid agencies such as the United Nations' World Food Program. There has also been record migration from Guatemala to the United States, as people have moved to find work. In addition, a state of emergency was declared in Honduras in 2019. Experts believe the droughts may continue as climate change has led to a disruption in weather cycles. Tropical cyclones—a major source of rain in Central America—are not landing as frequently in these areas. Central America has also become hotter and dryer over the past four decades.

A 2018 report commissioned by the World Bank states that in the next decades climate change will increase the numbers and the patterns of human "climate migrants." The report estimates 1.4 to 2.1 million people in Central America and Mexico will likely become climate migrants due to drought by 2050. Providing food aid to people left jobless and hungry by serious drought is one way to stem the flow of migrants. Some aid groups are already doing this, and Germany has agreed to give $1.1 million to the World Food Program for emergency food aid. The United Nations is also suggesting farmers switch to growing crops that are more drought tolerant, such as sorghum. Sorghum is a grain used to feed humans and animals.

A Guatemalan family, who entered the U.S. illegally as drought migrants from Guatemala, await a bus for their asylum hearing.

CHAPTER 5

Building Resilience

Thanks to scientific advances and technology, we are better prepared for heat waves and droughts than we have been in the past. Scientists also continue to look for new and better ways to predict these hazards further in advance, and new technologies are being developed to respond to and reduce their effects. But, it is clear that we can and need to do more.

Climate change is bringing challenges. It means heat waves and droughts are increasing in severity. Countries, cities, and individuals need to take responsibility for reducing their impacts on the planet. We need to build resilience to droughts and heat waves. We can do this through education, improved early warning systems, and infrastructure improvements such as the creation of green spaces and cool roofs. We can also encourage energy and water conservation and make personal efforts to do our part, such as not wasting water.

We should also work on developing an understanding that some areas of the world will be harder hit by heat waves and droughts, but we all need to be empathetic towards those who will suffer the most.

As far back as 1990, an Intergovernmental Panel on Climate Change (IPCC) stated that the greatest impact of climate change would be human migration. The panel thought millions of people would be forced to move from their homes and countries because of climatological disasters such as droughts, fires, or flooding. Decades later, we are just beginning to deal with the issues that surround severe drought and heat waves.

The Gobi Desert in Mongolia is the fastest growing desert on Earth. It swallows up over 1,300 square miles (3,370 km²) of land each year in a process called desertification.

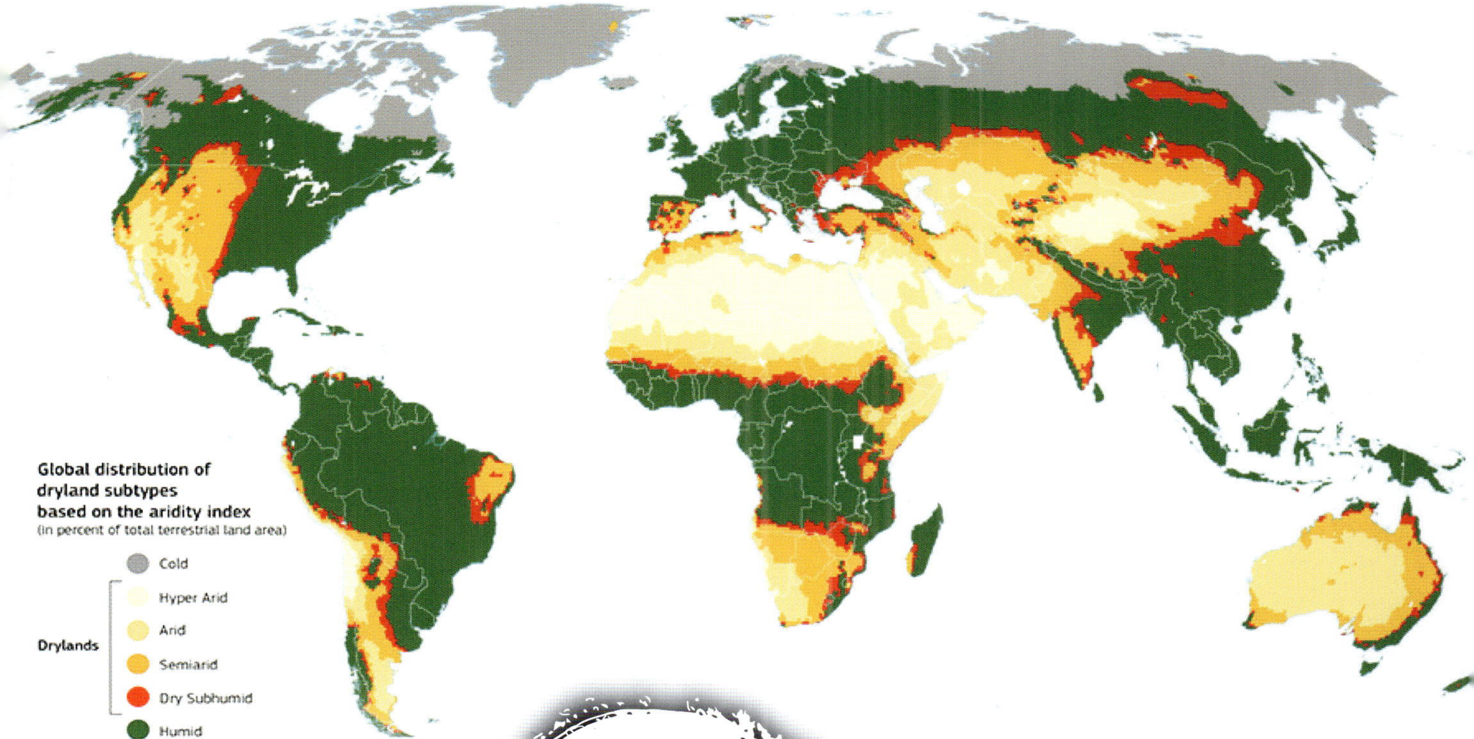

This map shows the global distribution of dry lands (green, orange, and pale yellow).

Ask Yourself This

Based on the information in this book, why do you think heat waves and droughts are such deadly disasters? Are there ways we can diminish the impact of these natural processes and help prevent disasters?

1. What have past heat waves and droughts taught countries and cities about preventing or reducing deaths and illnesses?
2. How does technology help us plan for disasters?
3. Climate change is causing more extreme weather around the world and is changing the way people live. What kind of actions can humans take to limit climate change or reduce its impact?

Bibliography

Introduction

Bhattacharya, Shaoni. "The 2003 European Heatwave Caused 35,000 Deaths." New Scientist. https://bit.ly/2TGjehz

"Causes and Effects of Drought." Earth Eclipse, April 24, 2016. https://bit.ly/2NIFrrG

"Increasingly, Droughts and Heat Waves Are Happening at the Same Time" Countdown to Armageddon. https://bit.ly/2tDpdZT

Kirkpatrick, Noel. "What Causes a Heat wave?" MNN, July 03, 2019. https://bit.ly/2RbnDaz

Kruszelnicki, Karl. "All You Need to Know about Heat waves." Cosmos, January 9, 2017. https://bit.ly/2ulcLOI

"Recent Australian Droughts May Be the Worst in 800 Years." University of Melbourne, Climate Energy College. https://bit.ly/2sLGeRo

"Where Do Heat Waves Occur in the World?" Reference. https://bit.ly/2T2Y37F

Chapter 1

Cappucci, Matthew. "Increasing humidity, driven in part of climate change, is making even modest heat waves unbearable." *The Washington Post*, August 13, 2019. https://wapo.st/2TGVF8w

"Causes of Drought: What's the Climate Connection?" Union of Concerned Scientists, n.d. https://bit.ly/2NI7o2T

"Education." National Drought Mitigation Center. https://bit.ly/3anc6wF

"Here are the places in Canada — yes, Canada — vulnerable to drought." *CBC News*, July 27, 2018. https://bit.ly/3avCDYq

Osborne, Hannah. "Decades without Rain: America Set for 'Megadroughts' Not Seen since Medieval Times." *Newsweek*, July 24, 2019. https://bit.ly/2NHcY5C

Perkins-Kirkpatrick, Sarah. "Winter Heatwaves Are Nice...as Extreme Weather Events Go." The Conversation, July 15, 2019. https://bit.ly/38rH7gS

"Quarter of World's Population Under Extreme Water Stress as Taps Run Dry, Drought Stares Them in Face." News18, August 26, 2019. https://bit.ly/3asrKa2

Wehner, Michael, Federico Castillo, and Dáithí Stone. "The Impact of Moisture and Temperature on Human Health in Heat Waves" Oxford Research Encyclopedias—Natural Hazard Science, November 14, 2017. https://bit.ly/2uhEbEW

Wolchover, Natalie. "What Is a Drought?" LiveScience, September 28, 2018. www.livescience.com/21469-drought-definition.html

Chapter 2

CGIAR Research Program on Water, Land, and Ecosystems. "Indian governments make the leap to drought relief for farmers using real-time data." https://bit.ly/2NJuh5X

Daley, Jason. "California's Drought Killed Almost 150 Million Trees." Smithsonian.com, July 10, 2019. https://bit.ly/2tAaqiJ

France-Presse, Agence. "'Unprecedented' Japan Heatwave Kills 65 People in a Week." *The Guardian*, July 24, 2018. https://bit.ly/2G6xXKN

Heat Index (definition). Accessed August 09, 2019. www.ambientweather.com/heind.html

"National Drought Mitigation Center." National Drought Mitigation Center. https://drought.unl.edu

Stokes, Elaisha. "The Drought That Preceded Syria's Civil War Was Likely the Worst in 900 Years." VICE News, March 03, 2016. https://bit.ly/30Fk13V

"Using the Excess Heat Factor (EHF) to Predict the Risk of Heat Related deaths." *Journal of Forensic and Legal Medicine*, January 17, 2013. Ward, Patrick Sawer; Victoria. "UK Weather:

Rails Overheat and Roads Melt as Temperatures Soar across Britain on Hottest Day of the Year." *The Telegraph,* June 29, 2019. https://bit.ly/2RGdTo6

Chapter 3

"After Drought." National Drought Mitigation Center. https://bit.ly/2tA4Z3j

"Can Droughts Be Predicted?" American Geosciences Institute, July 12, 2019. https://bit.ly/30CtGbi

"Case Study: Deadly Chicago Heat Wave of 1995." CUNY Graduate School of Journalism—Adapt NY. https://bit.ly/2RB8EFT

"CDC's Tracking Network in Action: Extreme Heat." Centers for Disease Control and Prevention. www.cdc.gov/features/trackingheat/index.html

EfficientGov Staff, "How Cities Worldwide Combat Drought & Conserve Water," EfficientGov, July 20, 2018. https://bit.ly/37cKqIR

Freedman, Andrew. "We May Soon Be Able to Predict Heat Waves 7 Weeks before They Happen." Mashable, March 28, 2016. https://bit.ly/2tAwgTe

Hao, Zengchao. "The Challenges of Drought Prediction." Eos, February 16, 2018. https://bit.ly/2TED6BE

Lieberman, Amy. "IFRC Issues New Guidelines to Help Cities Combat Deadly Heat Waves." Devex, July 17, 2019. https://bit.ly/2TKLXlp

"Researchers Use Space Tech to Predict Droughts Months in Advance." Water Source, February 6, 2019. https://bit.ly/2NLyzto

Sandru, Ovidiu. "Water Desalination Made Cheaper With Oasys Water's New Technology." The Green Optimistic, December 17, 2010. www.greenoptimistic.com/oasys-water-desalination-20101217

Schwartz, Ariel. "EarthRisk Can Predict Extreme Heat Waves Up To 40 Days In Advance." Fast Company, October 21, 2011. https://bit.ly/2tzmH6U

Toloo, Ghasem, Gerard Fitzgerald, Peter Aitken, Kenneth Verrall, and Shilu Tong. "Are Heat Warning Systems Effective?" *Environmental Health* 12, no. 1 (May 2013). https://doi.org/10.1186/1476-069x-12-27

U.S. Department of Commerce, and NOAA. "Heat Watch vs. Warning." National Weather Service, March 28, 2018. www.weather.gov/safety/heat-ww

Chapter 4

Aguilera, Jasmine. "A Deadly Heat Wave After the Hottest June On Record: How the Climate Crisis Is Creating 'a New Normal'" *Time*, https://bit.ly/38LOJLN

"Cities in the Face of Drought." *The Telegraph*, August 8, 2018. www.telegraph.co.uk/news/cities-in-the-face-of-drought

Elliott, Josh K. "Cities Get Hotter during Heat Waves. Here's How They Can Adapt for Climate Change." *Global News*, June 30, 2019. https://bit.ly/2RavXY8

"How 10 Western Cities Are Dealing with Water Scarcity and Drought." NPR. https://n.pr/2v0kP7A

Masters, Jeff. "Fifth Straight Year of Central American Drought Helping Drive Migration." Scientific American blogs, Dec. 23, 2019. https://bit.ly/30EFvxH

"NETAFIM History." Netafim. www.netafimuk.com/netafim-history

"Records Can Be Broken: Lessons from the Millennium Drought." International Water Association. https://bit.ly/38lVwvh

"Three Cities Innovate Solutions for Tackling Water Scarcity." Latest News. https://bit.ly/36f755U

Vong, Suong. "Protecting Climate Refugees is Crucial for the Future." Humanity In Action, May 2017. https://bit.ly/2NJfLLg

Learning More

Books

Furang, Kathy. *Everything Weather*. National Geographic Kids, 2012.

Howell, Izzi. *Climate Change Eco Facts*. Crabtree Publishing, 2020.

Reilly, Kathleen. *Natural Disasters, Investigate Earth's Most Destructive Forces*. Nomad Press, 2012.

Websites

Check out this site for information on weather and droughts. **www.weatherwizkids.com/?page_id=89**

The National Drought Mitigation Center has a website section for kids. **https://drought.unl.edu/Education/DroughtforKids.aspx**

News For Kids provides up-to-date news on heat waves. **https://newsforkids.net/articles/2018/07/23/worldwide-heat-wave**

Glossary

agriculture The science or practice of farming

civil war A war between organized groups in the same country

climate migrants People who move from their homes because they can no longer survive there because the climate has changed, bringing either drought, high water levels, or extreme weather events

climatological disaster Natural disasters related to extreme temperatures such as drought

concurrent Existing, happening, or being done at the same time

convection The transfer of heat by circulation or movement

desalination The process used to remove extra salt and minerals from water in order to get fresh water for drinking or irrigation

desertification The process in which land turns into desert, often as a result of drought

El Niño A climate pattern that brings warm water to the ocean surface in the tropical Pacific Ocean and that disrupts regional and global weather patterns

geological Relating to the study of Earth's physical structure

groundwater Water held underground in the soil or in holes and cracks in rocks

high pressure system A clockwise flow of high, sinking air that builds up behind a storm system

humidity A term used to describe how much water vapor is in the air. The more water there is in the air, the higher the humidity.

hydroelectric Energy produced from flowing water such as a waterfall

ice core An ice sample removed from an ice sheet or mountain glacier

indexes A list or record of something

infrastructure The basic facilities and systems needed for the functioning of an area such as roads, schools, hospitals, and housing

intensively To do something with a lot of effort in a short period of time

irrigation The watering of land to make it ready for agriculture

membrane A thin, soft, flexible outer layer

migrant A person who migrates

migration Temporary or permanent movement of people from one place to another

photosynthesis The process in which green plants use sunlight to make their own food

precipitation The liquid and solid water particles that fall from clouds and reach the ground

reservoir A large natural or artificial lake used to store water

resiliency The ability to spring back or recover quickly from something difficult

temperate climate Mild or moderate weather with winters that are not too cold and summers that are warm and sunny but not too hot

uninhabitable Unsuitable for living in

vegetation Word used to describe all plants and trees in an area as a whole

Index

About the Author

Rachel Seigel is an avid book enthusiast with over 15 years of experience working with schools and libraries matching books to readers. She is also the author of several non-fiction books for children. When she isn't writing or researching fun facts, she enjoys spending time with her boyfriend and her mischievous dog.